LIGHTING UP THE SKY

RAINBOWS

SETH KINGSTON

PowerKiDS press
New York

Published in 2021 by The Rosen Publishing Group, Inc.
29 East 21st Street, New York, NY 10010

First Edition

Portions of this work were originally authored by Kristen Rajczak and published as *Lightning*. All new material in this edition authored by Seth Kingston.

Editor: Elizabeth Krajnik
Book Design: Reann Nye

Photo Credits: Cover MichaelJust/iStock/Getty Images Plus/Getty Images; series art Zoteva/Shutterstock.com; series art Abstractor/Shutterstock.com; p. 5 Anke Wittkowski/EyeEm/Getty Images; p. 6 Mila Drumeva/Shutterstock.com; p. 7 Elena Breus/Shutterstock.com; p. 8 muratart/Shutterstock.com; p. 9 Peter Hermes Furian/Shutterstock.com; p. 11 Tiffani Nieusma/Moment/Getty Images; p. 13 © istockphoto/Joshua Moore; p. 14 © istockphoto/Adam-Springer; p. 15 Kevin Smith / Design Pics/Getty Images; p. 16 Lukassek/Shutterstock.com; p. 17 https://commons.wikimedia.org/wiki/File:Supernumerary_Rainbows.jpg; p. 19 john finney photography/Moment/Getty Images; p. 21 (top) orxy/Shutterstock.com; p. 21 (bottom) vlastas/Shutterstock.com; p. 22 Mike Mareen/Shutterstock.com.

Cataloging-in-Publication Data

Names: Kingston, Seth.
Title: Rainbows / Seth Kingston.
Description: New York : PowerKids Press, 2021. | Series: Lighting up the sky| Includes glossary and index.
Identifiers: ISBN 9781725318496 (pbk.) | ISBN 9781725318519 (library bound) | ISBN 9781725318502 (6 pack)
Subjects: LCSH: Rainbows–Juvenile literature. | Meteorological optics–Juvenile literature.
Classification: LCC QC976.R2 K56 2021 | DDC 551.567–dc23

Manufactured in the United States of America

CONTENTS

LOOK UP! 4
WHAT'S A PRISM? 6
THE VISIBLE SPECTRUM 8
FROM RAINDROPS TO RAINBOWS 10
SECONDARY RAINBOW. 12
COLORS OF THE RAINBOW 14
TYPES OF RAINBOWS. 16
RAINBOW WATCHING 18
POT OF GOLD 20
HOW DOES A RAINBOW FORM? 22
GLOSSARY. 23
INDEX. 24
WEBSITES 24

LOOK UP!

If you look up after a rainstorm, you might see an arc of colors reaching across the sky. Rainbows are one of the most beautiful **atmospheric** events. However, rainbows don't just occur after rain. They can form under other conditions too.

A rainbow occurs when light, which is made up of waves, passes through a **medium**, such as raindrops. Raindrops, prisms, and other mediums affect how light waves move. Light has to move through the raindrops in a special way for a rainbow to form. Keep reading to learn all about rainbows.

Even though rain may ruin your outdoor plans, the rainbow after a storm makes it worth it.

WHAT'S A PRISM?

A prism is a piece of glass or other clear material, such as plastic, that has two triangle-shaped **parallel** faces and three nonparallel long faces. When light enters through one of the prism's long faces, it separates into a rainbow. The light also refracts, or changes direction. Refraction happens because the light slows down when it enters the prism.

SEEING THE LIGHT

Even though a prism separates light into a rainbow, the colors of the rainbow can be combined to make white light again.

The amount the light refracts depends on how much the light's speed changes and the angle of the incident ray, which is the ray of light that entered the prism. When the light bends, it hits another face of the prism and is refracted again.

THE VISIBLE SPECTRUM

When white light enters a prism, it splits into a rainbow called the visible spectrum. This process is called dispersion. Each color on the visible spectrum has a different **frequency** and **wavelength**. Because of this, each color moves through the prism at a different speed. Violet has the highest frequency and shortest wavelength and red has the lowest frequency and the longest wavelength.

This image shows how white light is separated into the visible spectrum and how each color of the rainbow has a different angle of refraction based on its wavelength and frequency.

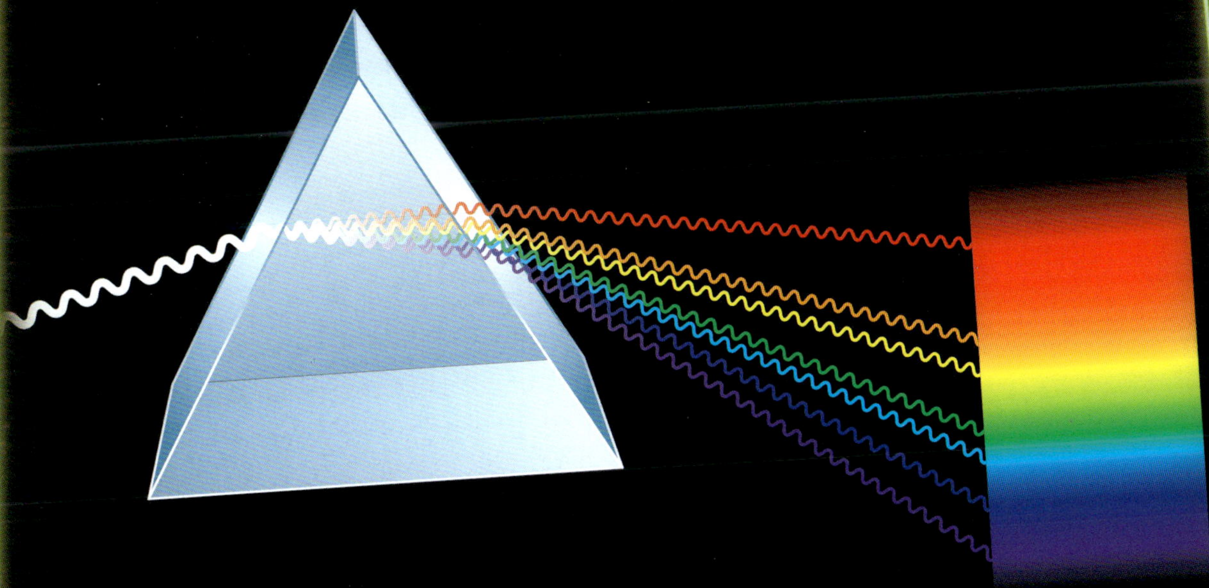

The longer the wavelength, the slower the light travels and the lesser the angle of refraction. The shorter the wavelength, the faster the light travels and the greater the angle of refraction. This means that red is refracted the least and violet is refracted the most.

FROM RAINDROPS TO RAINBOWS

After a rainstorm, raindrops act similar to a prism. However, unlike a prism, the curved inside of the raindrop causes the light to **reflect** back in the direction it came from. As light exits the raindrop, it's refracted again and disperses into the visible spectrum.

A rainbow is a very large way to see the visible spectrum. However, to create something this large takes more than just one raindrop. In fact, each raindrop only supplies one color to the rainbow. The color the raindrop supplies depends on the angle of refraction. Red light is refracted the least and is always on the outside of the rainbow.

SEEING THE LIGHT

Everyone sees a rainbow differently depending on where they're standing. Rainbows don't exist in one specific place. Many drops send light into your eye, which forms a cone of rays. This cone creates the image of the rainbow.

You always see rainbows in front of you. This means that the sun—the source of light needed to create a rainbow—will always be behind you when you see a rainbow.

SECONDARY RAINBOW

If you only see one rainbow, it's called a primary rainbow. If you see two rainbows, the second one is called the secondary rainbow. Primary rainbows have much clearer and darker colors. A secondary rainbow appears above the primary rainbow and has fainter and lighter colors. Secondary rainbows are often too faint to see.

Secondary rainbows form when light reflects twice within a raindrop. The reflections then exit the raindrop at different angles. The size of the raindrop must be just right for this to occur. The second reflection causes the secondary rainbow's colors to be opposite of those in the primary rainbow.

Secondary rainbows are wider than primary rainbows. Each color in a secondary rainbow is more widely spaced.

COLORS OF THE RAINBOW

Remembering the colors of the rainbow is easy with the help of a made-up name: Roy G. Biv. Each letter stands for one of the colors in the rainbow. They're in the order we see them in a primary rainbow. The colors we see are just the colors of the visible spectrum. However, more than 1 million colors make up a rainbow. Human eyes aren't able to see all these colors.

SEEING THE LIGHT

The space between a primary rainbow and a secondary rainbow is called Alexander's Dark Band. This area is darker than the sky below the primary rainbow and the sky above the secondary rainbow.

If you look out the window of an airplane, you might notice a rainbow circling the airplane's shadow on the clouds below. This **phenomenon** is often called **Pilot's** Glory, but it isn't actually a rainbow!

A rainbow is called a "bow" because its curved shape looks like a bow for shooting arrows. However, rainbows are actually circular—the curve of Earth just blocks all but an arc.

TYPES OF RAINBOWS

Rainbows can also occur at night. Bright moonlight, rather than sunlight, creates a **lunar** rainbow or moonbow. However, because even the brightest moonlight isn't as strong as sunlight, lunar rainbows are much fainter than those created by sunlight.

SEEING THE LIGHT

Lunar rainbows are rare. To create a lunar rainbow, the moon must be full or almost full and it must be raining opposite the moon. The sky must be dark and the moon must be at a certain height in the sky.

Supernumerary rainbows are faint rainbows inside a primary rainbow. Small, nearly the same sized raindrops cause supernumerary rainbows.

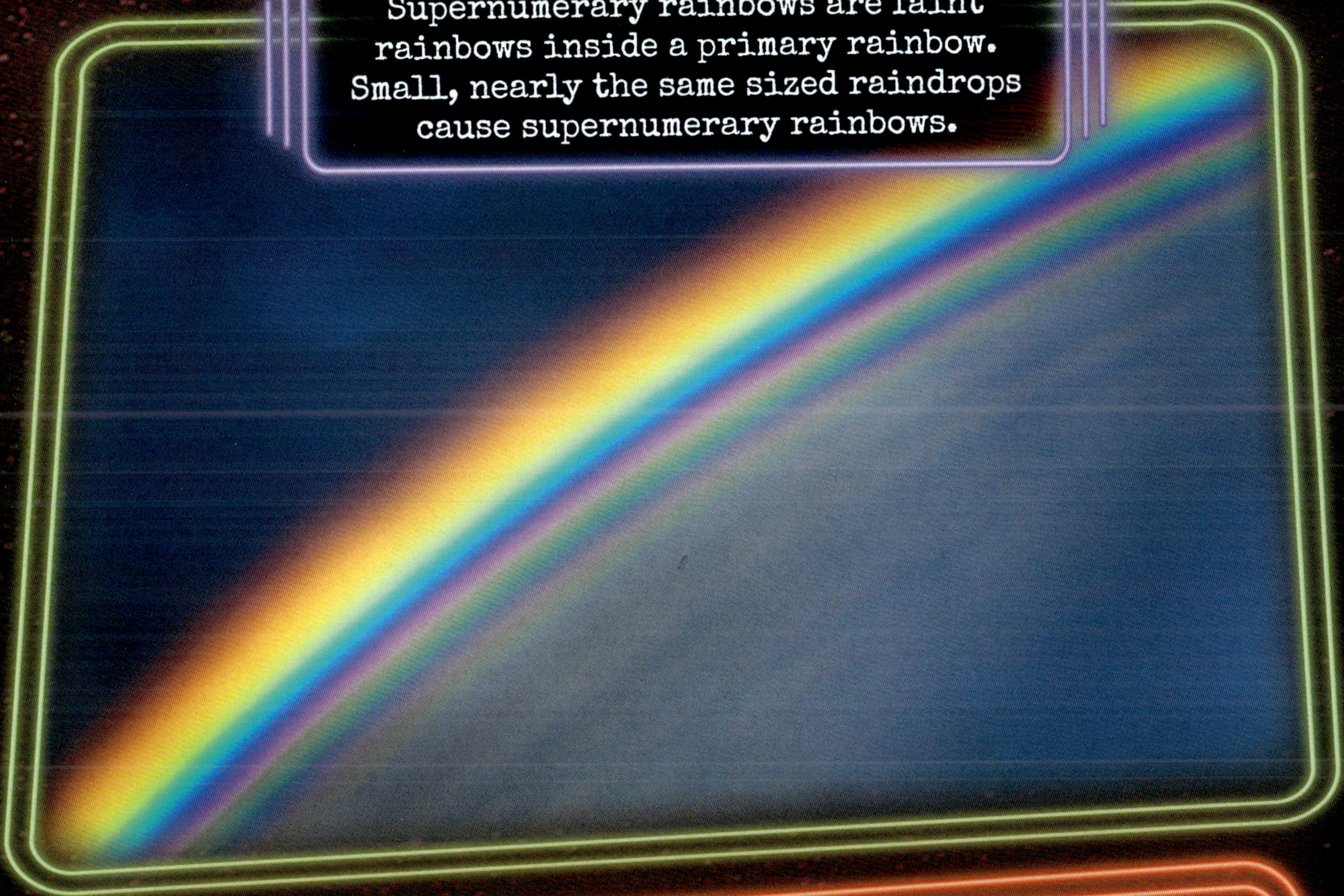

Have you ever seen a red rainbow? These rainbows happen most often at sunrise and sunset. The light at these times of the day travels long distances through the atmosphere. The blue and green colors are spread out the most due to their short wavelengths. This makes us see mostly red, orange, and yellow. Red rainbows are also taller than regular rainbows because the sun sits low in the sky.

RAINBOW WATCHING

We know that for a rainbow to form there must be raindrops and sunlight. That means it's more likely for rainbows to be seen in late spring, summer, and early fall than in winter, when rain turns to snow and ice.

On a rainy day, the best time to watch for a rainbow is when the sun is close to the **horizon**. The sun is closest to the horizon in the early morning and the late evening. At these times, the sun is low in the sky, which means the rainbow will have a higher arc.

The best time of day to watch for rainbows depends on where in the world you live and what time of year it is.

POT OF GOLD

It's a myth that you can find a pot of gold at the end of a rainbow. During the 17th century, people in Ireland would say that finding a pot of gold was as likely as finding the end of a rainbow.

Even then, people knew it was only possible to see part of a rainbow and that the rainbow moved away the closer they got to it. There's no end to a rainbow—or pot of gold—to find. Over time, the Irish saying changed, and people began saying there was a pot of gold at the end of the rainbow!

You've probably seen drawings and cartoons of pots of gold at the end of a rainbow. Have you ever tried to find the end of a rainbow to see if there's a pot of gold?

HOW DOES A RAINBOW FORM?

1. SUNLIGHT ENTERS RAINDROPS DURING A RAINSTORM.
2. SUNLIGHT IS REFRACTED.
3. SUNLIGHT REFLECTS OFF THE BACK OF THE RAINDROP.
4. SUNLIGHT IS REFRACTED AGAIN AS IT EXITS THE RAINDROP.
5. DISPERSION CAUSES SUNLIGHT TO SPLIT INTO THE VISIBLE SPECTRUM.

GLOSSARY

atmospheric: Of or relating to the atmosphere, which is the whole mass of air that surrounds Earth.

frequency: The number of waves of sound or energy that pass by a point every second.

horizon: The line where the earth or sea seems to meet the sky.

lunar: Of or relating to the moon.

medium: The substance in which something lives or acts.

parallel: Lying or moving in the same direction but always the same distance apart.

phenomenon: Something that can be observed and studied and that typically is unusual or hard to understand or explain fully.

pilot: A person who flies an airplane, helicopter, etc.

reflect: To move in one direction, hit a surface, and then quickly move in a different and usually opposite direction.

wavelength: The distance from one wave of energy to another as it's traveling from one point to another point.

INDEX

A
Alexander's Dark Band, 14
atmosphere, 4, 17

C
colors, 4, 6, 7, 8, 9, 10, 12, 13, 14, 17

D
dispersion, 8, 10, 22

F
frequencies, 8, 9

I
incident rays, 7
Ireland, 20

L
light, 4, 6, 7, 8, 9, 10, 11, 12, 17
lunar rainbows, 16

M
mediums, 4
myths, 20

P
Pilot's Glory, 15
prisms, 4, 6, 7, 8, 10

R
raindrops, 4, 10, 12, 17, 18, 22
refraction, 6, 7, 9, 10, 22

S
secondary rainbows, 12, 13, 14
sun, 11, 16, 17, 18, 22
supernumerary rainbows, 17

V
visible spectrum, 8, 9, 10, 14, 22

W
wavelengths, 8, 9, 17

WEBSITES

Due to the changing nature of Internet links, PowerKids Press has developed an online list of websites related to the subject of this book. This site is updated regularly. Please use this link to access the list:
www.powerkidslinks.com/lus/rainbows